MATH REVIEW TOOLKIT

Gary L. Long
Sharon L. Long

Fourth Edition
Ralph A. Burns

Upper Saddle River, NJ 07458

Project Manager: Kristen Kaiser
Senior Editor: Kent Porter Hamann
Editor in Chief: John Challice
Executive Managing Editor: Kathleen Schiaparelli
Assistant Managing Editor: Dinah Thong
Production Editor: Natasha Wolfe
Supplement Cover Management/Design: Paul Gourhan
Manufacturing Buyer: Ilene Kahn

 © 2003 by Pearson Education, Inc.
Pearson Education, Inc.
Upper Saddle River, NJ 07458

All rights reserved. No part of this book may be reproduced in any form or by any means, without permission in writing from the publisher.

The author and publisher of this book have used their best efforts in preparing this book. These efforts include the development, research, and testing of the theories and programs to determine their effectiveness. The author and publisher make no warranty of any kind, expressed or implied, with regard to these programs or the documentation contained in this book. The author and publisher shall not be liable in any event for incidental or consequential damages in connection with, or arising out of, the furnishing, performance, or use of these programs.

Printed in the United States of America

10 9 8 7 6 5 4

ISBN 0-13-033734-x

Pearson Education Ltd., *London*
Pearson Education Australia Pty. Ltd., *Sydney*
Pearson Education Singapore, Pte. Ltd.
Pearson Education North Asia Ltd., *Hong Kong*
Pearson Education Canada, Inc., *Toronto*
Pearson Educacíon de Mexico, S.A. de C.V.
Pearson Education—Japan, *Tokyo*
Pearson Education Malaysia, Pte. Ltd.
Pearson Education, *Upper Saddle River, New Jersey*

Contents

Acknowledgements v
Preface vii

1 Your Calculator and Its Functions 1
    1.1   Arithmetic Operations 1
    1.2   Trigonometric Operations 2
    1.3   Logarithmic Functions 2
    1.4   Choosing a Calculator 4

2 Skills for Chapter 3: Fundamental Measurements 6
    2.1   Significant Figures 6
    2.2   Scientific Notation 7
    2.3   Mean and Standard Deviation 9

3 Skills for Chapter 9: Chemical Quantities 10
    3.1   Molarity 10
    3.2   Dilution 12

4 Skills for Chapter 11: Stoichiometry 14

5 Skills for Chapter 12: Gases 16
    5.1   Gas Laws 16
    5.2   Ideal Gas Law 18
    5.3   Dalton's Law 19

6 Skills for Chapter 14: Solutions 22

7 Skills for Chapter 15: Reaction Rates and Chemical Equilibrium 23
    7.1   Calculations with Equilibrium Expressions 23
    7.2   Calculations of Equilibrium Concentrations 26

8	Skills for Chapter 16: Acids and Bases	*31*
	8.1 pH Calculations	*31*
	8.2 Titration	*34*
9	Skills for Chapter 18: Fundamentals of Nuclear Chemistry	*37*
10	Self Test for Math Skills	*39*
11	Chemistry and Writing	*42*
	11.1 The Scientific Notebook	*42*
	11.2 The Scientific Report	*44*
	11.3 Technical Writing	*46*
	11.4 A Notebook Example	*49*
12	Careers in Chemistry	*53*
	12.1 Materials Science	*53*
	12.2 Polymer Science	*54*
	12.3 Environmental Science and Technology	*54*
	12.4 Biochemistry and Biotechnology	*55*
	12.5 Medicinal Chemistry and Clinical Chemistry	*55*
	12.6 Forensic Science	*56*
	12.7 Radiochemistry	*57*
	12.8 Patent Attorneys and Patent Agents	*57*
	12.9 Chemical Education	*58*
	12.10 Chemical Information Careers	*58*
	12.11 Other Chemistry Careers	*59*
13	Study Aids for Chemistry	*60*
	13.1 Good Study Habits	*61*
	13.2 Reading Your Textbook	*62*
	13.3 Learning to Relax	*64*
	13.4 Conquering Tests	*66*
14	Answers for Math Skills Test	*71*

Acknowledgements

The authors (GLL and SDL) would like to acknowledge the contributions of Doris L. Lewis on the Sections 11 and 12 of this Toolkit on Chemistry and Writing, and Careers in Chemistry. Also acknowledged are the contributions of Timothy Smith and Diane Vukovich on Section 13 concerning Study Aids for Chemistry.

Preface   To The Student

A successful study of general chemistry requires you to use your skills of memorization, basic logic and mathematics. From my years in the classroom I have found that students who are unsure about their math skills generally do not do well on their exams. To put this statement into context, imagine that you holding a chemistry exam containing 20 multiple choice questions, with 15 of them involving the use of mathematical equations. If you have trouble with the first few questions involving math and have only an hour to complete the exam, any weakness in math skills may limit your performance on the exam, regardless of how many facts and figures you have memorized.

The math skills that a student needs to successfully complete general chemistry are basically what you have learned in high school; that is, algebra and trigonometry. It is the purpose of this booklet to guide you chapter by chapter in the mathematics that are used in the fourth edition text of **Fundamentals of Chemistry**. The math skills for each chapter are outlined and examples are worked in this booklet.

The case of a life sciences student planted the idea for this booklet. This young man was going around for the seventh time in general chemistry. He could never manage to pass the first test so he would drop out and wait until the next semester to try again. Fortunately, he came to me at the beginning of the course. He was desperate; he could not graduate until he passed chemistry. After speaking with him, I found his limitation was not inadequate math training, but a basic fear of math problems. He would stumble over the math problems on the exam. Over the course of the semester I met with the student and explained how to perform these calculations using the methods that are described in this booklet. The student successfully completed the course with a B average.

Although we cannot absolutely guarantee your success in chemistry with the use of this booklet, mastery of the material presented here will enhance your ability to work problems in general chemistry. Take a few moments and complete the "Self Test for Math Skills" at the back of this booklet. It will help you assess your skills. Use this guide to help you reinforce your math skills.

Chemistry is an exciting science that touches every aspect of our lives. It is our hope that this booklet may demystify the mathematics used in general chemistry and help you discover this excitement.

Gary L. Long
long@vt.edu

I will never forget my own struggle with chemistry at Wake Forest University. As a freshman, I had not yet developed productive study habits. I faced every new chemistry chapter with a certain dread. There is a tendency to procrastinate taking hold of something we fear or find distasteful, or so it was in my case. However, if you cope with chemistry in this manner, as I did, the stress of catching up and preparing for a test will cost you your well being until the test is over. The two grades of C that I made in freshman chemistry taught me this painful lesson. The point being, even if you dislike chemistry and are just taking it only to satisfy the requirements for your major, make your effort consistent.

Many college students have been forced to change their career dreams because they could not make it through chemistry. This unpleasantness does not give this science a good name nor does it give chemistry professors an easy grace at social functions when asked their occupation. Some of you may need to change your career option to a different discipline for which you are gifted. A very kind English professor pointed this out to me at a time when organic chemistry was causing me great distress. Time is too valuable to spend it in the wrong place. However, for those of you heading in the right direction, we do not want chemistry to be a stumbling block for your dreams.

We hope that your experience with chemistry will be a victorious one and that in some small way this math book will make a difference. Write us and tell us your stories as well as your triumphs.

Sharon D. Long
sdlong@vt.edu

Section 1

Your Calculator and Its Functions

A Peanuts cartoon from a few years ago displays Snoopy holding up a calculator and thinking, "I don't need to think, I have batteries..." This statement is far from reality in any study that requires the use of mathematics. Calculators are invaluable tools when working with long equations and involved calculations. What you can do in a few seconds on your calculator would have taken several minutes for a chemistry student 20 years. Even with this greater potential speed in problem solving, you still have to set up the equations before you use the calculator. Besides writing the equations down on paper, you may have to rearrange them so that you can use your calculator's functions to solve the problem. You will need to become acquainted with your calculator so you will know how to properly set up the problem. The mastery of the equations and your calculator's operation will allow you to get these "speedy" answers.

In your general chemistry course you will encounter problems using arithmetic operations (addition, subtraction, multiplication, and division), trigonometric operations (cos, sin, cos^{-1}, and sin^{-1}), and logarithmic operations (log, ln, 10^x, e^x, powers and roots). You will also deal with scientific notation in many of these problems. In the following sections, we will discuss these functions and how to use your calculator to solve problems in general chemistry.

1.1 Arithmetic Operations

These operations are generally straightforward. To add two numbers we enter the first number, press the [+] key, enter the second number and press the [=] key. The sum of the numbers should appear.

Some calculators may not follow this sequence because they are designed to use Reverse Polish Notation, RPN. These types of calculators do not have an [=] key. Calculations are performed by first entering the numbers into memory "stacks" on the calculator and then pressing the desired key.

1.2 Trigonometric Operations

You will encounter several equations in chemistry that will use the trigonometric operations of *sin* and *cos*. The process of determining the *sin* or *cos* of an angle is similar for conventional and RPN calculators. To determine the *sin* of 30°, enter 30 into the calculator and depress the [sin] key. The value of 0.50 should appear. Note: It is most important that your calculator is in the "degree" mode and not in "radian" or "grd" when performing these calculations.

Other calculations will require you to determine what degree value in *sin x* will correspond to a certain value, say $0.60 = sin\ x$. You can determine this x by using one of several keys on your calculator. Look for the [sin^{-1}] or [arcsin] key. Enter the value of 0.60 and then depress the [sin^{-1}], or the [arcsin] key. A value of 36.9° should appear on the display. The use of this function may require you to first depress a [2nd], [INV], [f], or [g] key.

1.3 Logarithmic Functions

These functions will include the use of *log*, *10^x*, *ln*, and *e^x*. Also to be discussed in this section are roots and powers.

In calculations involving logarithmic relationships, there are two different mathematical bases that are used. On your calculator you will find *base$_{10}$* logarithmic functions, [10^x] and [log], and *base$_e$* functions, [e^x] and [ln]. These logarithmic functions are not the same. *Base$_{10}$* math is structured around examining the properties of raising the number 10 to different powers, while *base$_e$* math is structured on using 2.71828.... You will find that many calculations, such as thermodynamic relationships, are defined using *log*, not *ln*. Part of this reasoning is historical, due to the fact that calculations involving $base_{10}$ math are much easier for people to work with than $base_e$ mathematics.

Many of these relationships were developed and refined by chemists before the advent of the computer and pocket calculator. For this reason the $base_{10}$ approach was favored. As an example, consider the calculation of the *log* of 1,000. Here, we must figure out to what power 10 must be raised to equal 1,000. We know from our algebra training that if we cube 10, we should obtain 1,000. The calculation of *ln* (1,000), which involves $base_e$, is another matter. To determine this value we must use the calculator. We first enter 1,000 and then press the [ln] key. The value of 6.908 should appear on the display.

The calculations of 10^x and e^x are sometimes referred to as anti-logs in calculator manuals. Most calculators have a [10^x] and a [e^x] key. Some models may require you to press the [INV] key to use an *antilog* function. In this scheme of thinking [e^x] = [INV] [ln]. To determine $e^{1.5}$, enter 1.5 into the calculator and depress the [e^x] or the [INV] [ln] key(s). A value of 4.5 should appear on the display.

Powers and roots are another important use of the calculator in chemistry. All calculators possess square root [$\sqrt{x}$] and square keys [x^2]. The use of these functions is straightforward. However, you may need to determine cubic (and higher) roots in your studies. Your calculator may have a [y^x] key that will enable you to determine these roots as well as to calculate a number raised to a power. To determine the roots with the [y^x] it is necessary to enter the root as the value of $1/x$. To find the cubic root of 77, enter 77 into the calculator (as the *y* value), enter 1/3 (as the *x* value), and depress the [y^x] key. You should see the value of 4.3 on the display.

Don't be discouraged if your calculator does not have a [y^x] key. If it has a [log] and a [10^x] or [INV] [ln] key(s), you can determine any root of a number. To calculate the cubic root of 77:

- enter 77 as the *y* value
- use the [log] button (1.8864 will be displayed)
- divide the result by 3, the xth root (0.6288 will be displayed)
- press the [10^x] button on the quotient (4.254 will be displayed)
- the result (4.3) is the cube root of 77

1.4 Choosing a Calculator

There are many excellent brands of calculators on the market available to the chemistry student. If you have not yet purchased a calculator, consider the following questions.

- Will you be pursuing a science, business, or liberal arts degree?
- Do you need a programmable or graphing calculator?

The calculator that you need in this course must be able to perform the basic functions in addition to trigonometric and logarithmic functions. It must support exponential notation. Prices of "scientific" calculators usually start at $15.

With regards to the first question, if you are planning to pursue a field of study in the liberal arts, the above mentioned calculator would well suit your needs. Science majors may wish to consider a calculator that contains a statistical analysis package and programmability in addition to the basic functions. Business majors would be well advised to purchase the basic "scientific" calculators. The calculations that you would use in later business courses will require you to use specific formulas and functions that are programmed into "business" calculators.

Programmability on a calculator is a wonderful feature that allows you to quickly solve repetitive calculations that are employed in science and engineering fields. Unfortunately, programmability adds to the cost of a calculator. Models that graph mathematical functions on a small LCD screen are also becoming popular, but at a higher cost. If you are operating on a restricted budget, choose a basic "scientific" model and upgrade later to a programmable or graphing model.

After you have answered these questions, examine the brands that would meet your calculating and financial needs. Examine the calculators for the size of the display and buttons (and their spacing) on the keypad. You want a calculator that has an easy to read display and that is easy to use. In order to put more "full feature functions" some manufacturers have down-sized the keys and put three to four functions on the same key. If the keypad legend is not well designed, these types of calculators can be difficult to use.

A final factor to consider is the power source. Solar powered calculators are great while working in well-lighted areas. Unfortunately, these calculators may not work well in some large lecture halls because of insufficient room lighting. Some calculators have dual power sources: solar and battery. With this type of design, your batteries will power the calculator if the light levels are too weak. (This is especially good to know before you take your first chemistry test.)

Your calculator can be a valuable tool to you in your study of general chemistry. Take the time to acquaint yourself with its capabilities.

Section 2

Skills for Chapter 3: Fundamental Measurements

2.1 Significant Figures

Our calculators normally display six to nine digits when performing arithmetic functions. While it may seem advantageous to have such large numbers when performing chemical computations, not all of this information may be significant because of error associated with the data.

To put this into context, consider the data that you record in the laboratory. Every measurement made is subject to error. The level of this error (or certainty) depends on the instruments used in the measurement and the skill of the person performing the operations. Even if we can eliminate the systematic errors (*i.e.* miscalibration of the pan balance) we will still encounter random errors in lab. These random errors will determine the accuracy of our measurement.

Rather than stating the error with each number a chemist may use in a calculation, the practice of using "significant digits" in calculations is employed. The significant digits of a number can be thought of as those digits in the number that do not change when the uncertainty is factored in. (The rules used for determining the number of significant digits in an expressed value are listed in Section 3.8 of your textbook.) As an example of the effect of uncertainty on the number of significant digits, consider a food sample, which was found to have 1.33 mg of Ca per serving with an error of 0.1 mg for the determination. Based on the uncertainty we should only say that the sample contains 1.3 mg of Ca. The number has only two significant digits. It is of no importance to attempt to convey that the calculation indicated a second 3 in the decimal. The error made this digit "insignificant." For practice in significant digits and rounding off nonsignificant digits, see Example

3.16, Exercise 3.16 and Problems 3.41 and 3.42. Problems 3.43 and 3.44 deal with the rounding of numbers.

You will be asked in this course to use significant digits in the estimation of answers to your calculations. As an example, consider the problem of determining the circumference of a cylinder. With a simple ruler you could determine the diameter to be 2.5 inches. From the relationship of $C = \pi \times d$, you could use your calculator to determine the circumference, C. The display of the calculator would read (7.853981...). What value do you report? What is significant?

The basic rule is that you report your answer using the least number of significant digits. In the multiplicative operation above, the diameter was only reported to two significant digits. The answer should be reported as 7.9 in. See Example 3.17, Exercise 3.17 and Problems 3.47(b,c) and 3.48(b,c) for more work with multiplicative operations and significant digits.

Rules for addition and subtraction are slightly different. Consider adding a 0.001 g weight to an object resting on a pan balance and indicating a value of 23.2 g in mass. The addition of 0.001 to 23.2 would yield a theoretical answer of 23.201 g. However, if the pan balance used had an error of 0.1 g, the addition of 0.001 g to this weight would be insignificant. The uncertainty will limit our ability to express accurately the sum of the combined weights. See Problems 3.47(a) and 3.48(b) for work with significant figures for addition and subtraction.

2.2 Scientific Notation

Scientific notation is a method used to express very large and very small numbers on your calculator. These numbers are expressed by multiplying the significant portion of the number by a multiplier based on 10^x.

To use the scientific notation feature on your calculator, you will use a [sci], [exp] or [EE] function. Consider the process of entering 1.6×10^{-17}. First enter the significant portion of 1.6 into the display. Next press the [sci] or equivalent key and enter the number 17. You will need to change the sign of the power of 17 (which should appear in the right hand portion of the display), so press the [+/-] key to cause the power to read -17. Depending on the type of calculator you have, you may need to press the [=] key to end the sequence.

You will also be able to convert from regular notation to scientific notation on your calculator. If the number 1,000,000 was in the display of your calculator, you could depress the [sci] then [=] keys to cause the number to appear in scientific notation (1.000 × 10^6) on the display.

An example of scientific notation involves the determination of the frequency of a photon from its wavelength (See Chapter 5, page 122 of your textbook). For instance, if a HeNe laser has a wavelength of 632.8 nm, what is the frequency of the photon? The answer is calculated using $c=v\lambda$, where c is the speed of light (a constant), v is the frequency (in Hertz) and λ is the wavelength in meters. We can quickly estimate the frequency by expressing all numbers in scientific notation and setting up the problem. Using this, we can write:

$$c = 2.998 \times 10^8 \text{ m/s}$$
$$\lambda = 632.8 \text{ nm or } 632.8 \times 10^{-9} \text{ m}$$

or 6.328×10^{-7} m

The problem can be set up to solve for v as:

$$v = \frac{c}{\lambda}$$

$$v = \frac{2.998 \times 10^8 \, m/s}{6.328 \times 10^{-7} \, m}$$

$$v = 4.738 \times 10^{14} \, Hz$$

Before using the calculator, we can get a rough idea of the magnitude of the number. The exponential portion of the calculation shows a +8 in the numerator and a -7 in the denominator. Using our algebra skills, we can determine the magnitude of the exponent is +15. The non-exponential term of the equation can be reduced to 3/6, which is 1/2 (or

0.5). Hence, our rough guess of the answer is 0.5×10^{15} Hz, or 5×10^{14} Hz.

The calculator allows us to find the answer of 4.738×10^{14} Hz, based on 4 significant digits. This answer is close to our estimate. More examples of scientific notation can be found in Example 3.18, Exercise 3.18, and Problems 3.45 and 3.46.

2.3 Mean and Standard Deviation

Chapter 3.7 of your text discusses the uncertainty that can be associated with measurements in the laboratory. This uncertainty is often not evident from just looking at the data. Instead, the uncertainty is often calculated based upon all the data you have available. Often, these calculations involved the determination of the mean and standard deviation of the data. These two values are helpful in determining the accuracy and precision of your measurements.

If your calculator has a statistical analysis package, you may be able to calculate the mean ($\bar{x}$) and standard deviation (s) of a data set by simply entering the data into calculator and pressing the appropriate function key. As an example, let us say you titrated five equal volumes of an unknown acid with a standard base. You have dispensed the following volume of titrant in the 5 runs: 43.80 mL, 43.30 mL, 44.10 mL, 43.90 mL and 43.70 mL. We would first enter the five values into the memory of the calculator. After the final entry, we would depress the button to give us the mean, $\bar{x}$. The value of 43.76 should appear. Next, we would calculate the standard deviation, s, by pressing the appropriate buttons (s, s_x, or σ). A value of 0.30 should appear.

Based on your titrating skills, you could say you dispensed 43.76 +/- 0.30 mL per titration in this data set. Considering the error in the data (0.30 mL) expressing a value beyond the tenths of mL (0.1 mL) would not be significant. The correct answer should be 43.8 ± 0.3 mL, where only three significant digits are used.

Section

 **Skills for Chapter 9: Chemical Quantities**

Goals: To calculate concentration and dilutions.
Skills: Multiplication and division.

3.1 Molarity

Molarity, M, is the term that expresses the mol quantity of a dissolved substance (solute) per liter of solution. A solution that is 1.00 M NaCl contains 1.00 mol of NaCl per 1.00 L of solution. This same solution contains 58.4 g of NaCl per 1.00 L of solution.

Calculations that involve molarity will usually begin with statement of the gram quantities of solute dissolved in a quantity (mL or L) of solution. By definition,

$$M = \frac{\text{mol of solute}}{\text{L of solution}} = \frac{\left(\frac{\text{g of solute}}{\text{MW of solute}}\right)}{\text{L of solution}}$$

For instance, to calculate the molarity of a solution that is 10.0 g of NaCl in 400 mL of water, we would write:

$$M = \frac{\left(\frac{10.0 \text{ g}}{58.4 \text{ g/mol}}\right)}{0.400 \text{ L}}$$

we will first solve the numerator of the problem for the mol amount of NaCl present in the solution. If we divide 10.0 g by 58.4 g/mol, we should obtain a value of 0.171 mol NaCl. Substituting into the equation

$$M = \frac{0.171 \text{ mol}}{0.400 \text{ L}}$$

$$M = 0.428 \text{ mol NaCl / L of solution}$$

Note: It is most important to express the volume of solution only in L.

You may find it necessary in your work to determine what amount of solute, expressed in grams, would result in a certain molarity of solution. Consider what gram amount of NaCl is necessary to create a 0.60 M solution that is 750 mL in volume. First, let's fill the given information into the molarity expression.

$$0.60 \text{ M} = \frac{\left(\dfrac{x \text{ g}}{58.4 \text{ g / mol}}\right)}{0.750 \text{ L}}$$

To solve for x, we must cross multiply several factors in order to arrange the equation for x.

$$(0.60 \text{ M})(0.750 \text{ L}) = \frac{x \text{ g}}{58.4 \text{ g/mol}}$$

$$(0.60 \text{ mol/L})(0.750 \text{ L})(58.4 \text{ g/mol}) = x \text{ g}$$

$$x \text{ g} = 26.3 \text{ g of NaCl}$$

Note that all units on the left hand side of the equation reduced to g (remember that M is in mol/L). By taking time with your cross multiplication step, the problem can be solved. This example is somewhat similar to Example 9.13, Exercise 9.13 and Problems 9.47-9.50 of your textbook.

Another use of the molarity equation involves the determination of the volume of solution needed to deliver a specific quantity of a chemical reagent. See Example 9.14, Exercise 9.14 and Problems 9.51-9.54.

Additional work with molarity calculations is found in Chapter 14 of your text. See Examples and Exercises 14.7 and 14.8, and Problems 14.41-14.46.

3.2 Dilution

Dilution involves reducing the concentration of a solution by adding more solvent. Typically, a volume of solution is taken and added to a volume of solvent. Consider the dilution of a 3.00 M NaCl stock solution by pipetting 10.0 mL into a 1.000 L flask and filling the flask to the calibration mark with pure water. Here we are adding roughly 990 mL of water to the solution. The concentration of the diluted solution is calculated as:

$$M_S \times V_S = M_D \times V_D$$

$$M_D = M_S \times \frac{V_S}{V_D}$$

where M_D is the final concentration of the diluted solution, M_S is the initial concentration of the concentrated solution, V_D is the final concentration of the diluted solution, and V_S is the initial concentration of the solution. For our problem, M_S = 3.00 M NaCl, V_S = 10.0 mL, and V_D = 1.000 L (or 1,000 mL). Substituting these numbers into the equation, we write:

$$M_D = 3.00 \text{ M} \times \frac{10.0 \text{ mL}}{1000 \text{ mL}}$$

$$M_D = 0.0300 \text{ M}$$

Note that the volume units in a dilution calculation must be the same no matter how they are stated in the problem. Use dimensional analysis (Chapter 3.3) to check your units before reporting your answers.

The dilution equation can also be used to solve for V_S, if the other variables of the equation are known. For instance, let's consider what volume (V_S) of a 16.0 M solution of HNO_3 (M_S) must be added to beaker and diluted to 2.25 L (V_D) in order to produce a final solution of 0.10 M HNO_3 (M_D). Rearrangement of the dilution equation by cross-multiplication and division yields,

$$V_S = V_D \times \frac{M_D}{M_S}$$

$$V_S = 2.25 \text{ L} \times \frac{0.10 \text{ M}}{16.0 \text{ M}}$$

from which V_S can be found as 0.0141 L (or 14.1 mL). Example 9.15 is somewhat similar to this problem. Also see Exercise 9.15 and Problems 9.55-9.58 in your textbook for more practice in calculating dilution volumes. Additional work with dilution is found in your text in a later chapter. See Example 14.12, Exercise 14.12 and Problems 14.65-14.70.

Section 4

Skills for Chapter 11: Stoichiometry

Goals: To calculate yields.
Skills: Multiplication and division.

Part of your studies dealing with chemicals and their reactivity will be to determine the yield of a reaction. From your readings in Chapter 11 concerning stoichiometry, if we write the equation,

$$A + B \rightarrow C$$

we understand if 1 mol of A and 1 mol of B are allowed to react under favorable conditions, 1 mol of C should be produced. The *theoretical yield* of the reaction is 1 mol of C. This value could be expressed in grams instead of moles by multiplying the theoretical yield of 1 mol by the molar mass of the compound C.

Note that yield depends on the stoichiometry of the reaction. If the reaction were instead,

$$2A \rightarrow B$$

and 1 mol of A was placed in the reaction vessel, the theoretical yield would be 0.5 mol of B. Using the unit analysis problem solving method (Chapter 3.3 of your text) the problem could be solved.

$$x \text{ mol} = 1 \text{ mol A} \times \frac{1 \text{mol B}}{2 \text{mol A}}$$

$$x \text{ mol} = 0.5 \text{ mol B}$$

In practice, the mol quantity of product obtained from the reaction is less than the theoretical value. The amount of product obtained is termed the *actual yield*.

A method in which the efficiency of a reaction is gauged is the calculation of the *percent yield*. It is defined as

$$percent\ yield = \frac{actual\ yield}{theoretical\ yield} \times 100\%$$

For the first reaction, if we had obtained 0.8 mol of C, the % yield would be

$$percent\ yield = \frac{0.8\ mol}{1\ mol} \times 100\%$$

$$percent\ yield = 80\%$$

The percent yield calculations can also be performed using the grams of actual product obtained as compared to the theoretical gram yield. Example 11.7 and Exercise 11.7 involve the calculation of theoretical yield and percent yield. See Problems 11.21-11.29 for more work with the calculations and use of yields.

In many cases, the scientist may wish to predict the actual yield for a well-characterized reaction. For instance, if the percent yield of a synthetic reaction is 90%, the actual yield may be found by multiplying the percent yield by the theoretical yield. (Remember to divide the percent yield by 100% in order to remove the % unit from the expression.)

Section 5

Skills for Chapter 12: Gases

Goals: To perform calculations with gas laws and the ideal gas law.
Skills: Cross multiplication and division.

5.1 Gas Laws

In this section of your studies, you will be exploring relationships between pressure, P, volume, V, temperature, T, and mol quantities of gases, n. Your text will introduce you to Boyle's law, Charles's law, Gay-Lussac's law, Avogadro's law, and the combined gas law.

You must exercise care in two areas to avoid mistakes when using these laws. The first area is temperature: all temperatures must be expressed in Kelvin (K) even if the data is given to you in °C. Remember, the relationship is K = °C + 273.15. The second area is to rearrange the equation so that the left hand side only contains the term that you are seeking. All the given data would then appear on the right hand side of the equation.

Consider the use of Charles's law in finding the volume that a gas in a balloon would occupy if it was heated from 25°C to 75°C. (This law is discussed in Chapter 12.5 of your textbook.) The initial volume of the gas (V_1) at 25°C is 1.00 L. First, let's set up the equation:

$$\frac{V_1}{T_1} = \frac{V_2}{T_2}$$

From the above information we can write:

V_1 = 1.00 L
T_1 = 25°C, which is 298 K

$$V_2 = ?$$
$$T_2 = 75°C, \text{ which is } 348 \text{ K}$$

To solve for V_2, we write:

$$\frac{V_1 T_2}{T_1} = V_2 \qquad \text{(cross multiplying } T_2\text{)}$$

$$V_2 = \frac{V_1 T_2}{T_1} \qquad \text{(switching sides)}$$

By plugging in the data, we obtain:

$$V_2 = \frac{1.00 \text{L} \times 348 \text{K}}{298 \text{K}}$$

$$V_2 = 1.17 \text{ L}$$

When dealing with Boyle's, Charles's, Gay-Lussac's law, or the combined gas law, you will often find a great deal of information in the problem. Follow these steps to solve such problems:

1) write the gas law equation to be used.
2) sort the given data (as set 1 or set 2).
3) rearrange the equation to solve for the unknown term.
4) enter the data and solve.

Problems with Boyle's Law are found in Example 12.1, Exercises 12.1, and Problems 12.27-12.30. Charles's law problems are found in Example 12.3, Exercise 12.3, and Problems 12.35-12.38. Work with Gay-Lussac's Law is found in Example 12.4, Exercise 12.4, and Problems 12.42-12.44, while Avogadro's law is covered in Example 12.8, Exercise 12.8, and Problems 12.53-12.56. The combined gas law is addressed in Example 12.7, Exercise 12.7, and Problems 12.45-12.50.

5.2 Ideal Gas Law

Another law that you will use is the ideal gas law, $PV = nRT$. It is described in Chapter 12.11 of your textbook. Here, n is the number of moles of the gas and R is the gas constant [(0.0821 L·atm)/(K·mol)]. This gas law will allow you to work any of the previous problems that relate the P, V, and T of a gas to the mole quantity of the gas. For instance, what volume would 1.00 g of CO_2 occupy at 1.00 atm of pressure and a temperature of 23°C? Begin this problem by writing the equation.

$$PV = nRT$$

We were not given n in the original problem, but we can calculate it by using the equation $n = g/M_M$ (where M_M is the molar mass of the gas). Here,

$$n = \frac{1.00 \text{ g}}{44.0 \text{ g/mol } CO_2}$$

and is equal to 0.0227 mol. With this information we can write:

$P = 1.00$ atm
$V = ?$
$n = 0.0227$ mol
$R = 0.0821$ (L·atm)/(K·mol)
$T = 23°C$, which must be expressed as 296 K

Rearranging the equation to solve for V, we write:

$$V = \frac{nRT}{P}$$

Plugging in the data:

$$V = \frac{0.0227 \text{ mol} \times 0.0821 \text{ (L·atm)/(K·mol)} \times 296 \text{ K}}{1.00 \text{ atm}}$$

$$V = 0.551 \text{ L}$$

Example and Exercise 12.10, and Problems 12.71 and 12.72 make use of the ideal gas law for finding P, V, n, or T of gases.

Another important use of the ideal gas law involves the determination of the molar mass of a gas. The molar mass (M_M) of a gas is the weight of one mole of gas. It can be determined by replacing the n term in the ideal gas law with g/M_M, where g represents the grams of the gas. Example 12.11, Exercise 12.11, and Problems 12.73 and 12.74 shows the use of M_M in the ideal gas law.

A final application of the ideal gas law is the determination of the density of a gas. The ideal gas law can be quickly rearranged to determine grams of gas per liter. This is illustrated in Example 12.9, Exercise 12.9, and Problems 12.63-12.66.

5.3 Dalton's Law

The pressure of a gaseous mixture is dependent on the partial pressure exerted by each gas. Dalton's law of partial pressures describes this relationship and is expressed as $P_1 = X_1 P_{total}$, where P_1 is the pressure of gas_1, X_1 is the mole fraction of gas_1, and P_{total} is the total pressure of the gas. To illustrate this law, consider a container where 1.0 mol of gas_1 is mixed with 3.0 mols of gas_2. The mole fraction of gas_1 is calculated as follows:

$$X_1 = \frac{1.0 \text{ mole of gas}_1}{4.0 \text{ moles of gas total}}$$

$$X_1 = 0.25$$

similarly,

$$X_2 = \frac{3.0 \text{ moles of gas}_2}{4.0 \text{ moles of gas total}}$$

$$X_2 = 0.75$$

IMPORTANT: The sums of all the mole fractions should always add up to 1. Since the mole fraction is a simple ratio, it has no units.

In working the problem, we are told the pressure in the container is 2.00 atm. What then is the partial pressure of gas_1? The equation is:

$$P_1 = X_1 P_{total}$$

The terms to use in the equation are:

$X_1 = 0.25$ (calculated above)
$P_{total} = 2.00$ atm (given in the problem)

Now by substituting into the equation we obtain,

$$P_1 = 0.25 \times 2.00 \text{ atm}$$
$$P_1 = 0.50 \text{ atm}$$

By similar reasoning we could determine that $P_2 = 1.50$ atm. Also since the gas mixture consisted of only 2 gases, if the pressure of gas_1 is 0.50 atm, and the total pressure of *both* gases is 2.00 atm, by the difference we can find the pressure of gas_2 to be 1.50 atm.

Example 12.12 and Exercise 12.12(b) introduce you to the use of partial pressure calculations for the determination of the pressure of a gas that is produced in chemical reaction. In these examples

$$P_{gas\ collected} = P_{total} - P_{water\ vapor}$$

The knowledge of the T of the system allows $P_{water\ vapor}$ to be found (from Table 12.2 of your text). P_{total} is normally given from the reading of a pressure gauge. The simple subtraction of $P_{water\ vapor}$ from P_{total}

gives you $P_{\text{gas collected}}$. This value can then be used with the ideal gas law to find the moles of gas that were produced in a chemical reaction. See Problems 12.79 and 12.80 for work with partial pressures.

Section 6

Skills for Chapter 14: Solutions

Goals: To determine concentrations of solutions
Skills: Cross multiplication and division

The calculations in this chapter involve the use of the concentration expressions molarity, % by volume and % by mass. Also discussed are the units ppm (parts per million) and ppb (parts per billion) for describing dilute solutions.

Molarity was introduced earlier in Chapter 9 of your text. In Chapter 14, work with molarity calculations is found in Examples 14.7 and 14.8, Exercises 14.7 and 14.8, and Problems 14.41-14.46.

The unit of % by volume is addressed in Example 14.9, Exercises 14.9 and Problems 14.47-14.52. Another unit that is based on percent is % by mass. These calculations are shown in Example 14.10, Exercises 14.10 and Problems 14.53-14.56.

Concentration units for dilute solutions (ppm and ppb) are used in Example 14.11 and Exercise 14.11. Problems are 14.57-14.64.

Section

Skills for Chapter 15: Reaction Rates and Chemical Equilibrium

Goals: To calculate concentrations of products and reactants at equilibrium, and equilibrium constants.
Skills: Cross multiplication, powers, and roots.

Chapter 15 of your text introduces you to the concepts of chemical equilibrium. It is based upon having a known stoichiometric chemical reaction in which the equilibrium expression can be written. With this information, it is possible to predict the yield of a chemical reaction as well as determining the value of a equilibrium constant.

7.1 Calculations with Equilibrium Expressions

As an example of calculations you will use in your study of equilibrium, consider the chemical reaction $A + B \rightarrow 2C$. The equilibrium expression for the reaction is:

$$K_{eq} = \frac{[C]^2}{[A][B]}$$

If you are given the equilibrium concentrations of A, B, and C, the equilibrium constant can be determined by entering these concentrations into the equation. For instance, if [A] = 0.10 M, [B] = 0.20 M, and [C] = 0.50 M, we can write:

$$K_{eq} = \frac{[0.50]^2}{[0.10][0.20]}$$

and

$$K_{eq} = 13 \quad \text{(with only 2 significant figures)}$$

A variation to this problem may involve the determination of [C] if [A], [B], and K_c are given to you. If K_c = 13 with [A] = 0.50 M and [B] = 0.25, let's arrange the above equation to solve for [C].

$$K_c = \frac{[C]^2}{[0.50][0.25]}$$

Next, each side of the equation is multiplied by the denominator, [0.50][0.25]. This act leaves only the term $[C]^2$ on the right hand side of the equation.

$$13 \times ([0.50][0.25]) = [C]^2$$

By rearranging and multiplying the numbers we generate,

$$\sqrt{[C]^2} = \sqrt{1.6}$$

$$[C] = 1.3$$

The equilibrium concentration of C is 1.3 M.

There may be times in solving the equilibrium problems where the determination of a concentration involves taking the third or fourth root of the expression. For instance, let's look at the Haber process (page 460 of your textbook):

$$N_{2(g)} + 3\ H_{2(g)} \leftrightarrow 2\ NH_{3(g)}$$

The equilibrium expression for this reaction is:

$$K_c = \frac{[NH_3]^2}{[N_2][H_2]^3}$$

To solve for K_c, we must raise the equilibrium concentration to the power of 3 for hydrogen, while squaring the ammonia concentration.

If, instead, you wish to solve for the hydrogen concentration using a given K_c value and known equilibrium concentration of ammonia and nitrogen, the equation must be rearranged.

$$[H_2]^3 = \frac{[NH_3]^2}{K_c[N_2]}$$

If K_c =9.52, [NH$_3$] = 0.030 M, and [N$_2$] = 0.050, what is [H$_2$]? Plugging in the numbers, we can write:

$$[H_2]^3 = \frac{(0.030)^2}{9.52 \times (0.050)}$$

$$[H_2]^3 = 0.00189$$

$$[H_2] = (0.00189)^{1/3}$$

$$[H_2] = 0.12 \text{ M} \quad \text{using the } y^x \text{ function}$$

In order to work any of these types of problems, it is most important that you be able to write the correct equilibrium constant expression for the reaction. If you are in doubt as to how to construct the expression, look back at Chapter 15.6 of your textbook.

7.2 Calculation of Equilibrium Concentrations

A second use of the equilibrium expression involves the determination of the concentrations of the products and the reactants at equilibrium. If the stoichiometry of the reaction is known (allowing the equilibrium expression to be written) and K_c is known, then the equilibrium concentrations may be found.

However, the solution to this problem is not always straightforward. In most cases, we only know the *initial* concentrations of the products and the reactants; we don't know the *equilibrium* concentrations. To calculate these equilibrium concentrations we must figure out how much the initial concentrations of products and reactants change, based upon the relationship defined by the equilibrium expression.

To help explain how such a solution to this problem is found, we will introduce here the "equilibrium checkbook". This teaching tool will allow us to quickly determine the concentrations of all equilibrium products and reactants. The checkbook concept is patterned very much like the checkbook that you are maintaining with a bank in your town. Just as keeping a good record of all deposits and drafts to the account enables you to balance the account, the equilibrium concentration of products and reactants can be readily determined with this method.

To see how the checkbook works, let's return to the "alphabet" reaction of Section 7.1 of this toolkit: $A + B \rightarrow 2C$. This balanced equation shows that 2 moles of product C are produced when 1 mole of A reacts with 1 mole of B. The equilibrium expression is then written as:

$$K_c = \frac{[C]^2}{[A][B]}$$

For the purpose of this example, let's say that $K_c = 0.0100$. The initial concentrations of both A and B are 0.500 M. The initial concentration of C is zero.

Because there is no C present, we can easily see that the reaction will proceed toward the right; that is, some C will be produced.

However, there are times when you will be asked to perform calculations such as these when the concentration of the 'products' (those written on the right-hand side of the equation) are not zero. To determine which direction a reaction will take (*e.g.* to the right or to the left), you should calculate the reaction quotient, Q. The formula seems identical to the equilibrium expression, but note that the initial concentrations, as noted by ( ) brackets, of the products and reactants are used rather than the equilibrium concentrations, as noted by [ ] brackets.

$$Q = \frac{(C)^2}{(A)(B)}$$

By comparing Q to K_c, the direction of the reaction can be determined.

In the sample problem for demonstrating the equilibrium checkbook, we have the task of finding [C] from the equilibrium equation. The setup of the checkbook is shown below:

Species	A	B	2 C
initial			
Δ			
equilibrium			

In the top row (labeled species), the reactants and products of the chemical equation are written. The stoichiometry of the equations is also entered in the top row of each column. Note that the double line separates the reactants and products and serves in place of the ↔ symbol. The other rows are labeled: initial, Δ, and equilibrium. These entries will pertain to how the concentration of each reactant and product change during the course of the reaction.

With the given information, we can set up the entries in the equilibrium checkbook. Our initial concentrations are entered in the initial row under the proper column of each reactant and product.

Species	A	B	2 C
initial	0.500 M	0.500 M	0 M
Δ			
equilibrium			

The next stage is to determine the changes (withdrawals and deposits) to each account. From the stoichiometry of the reaction, every 1 mol of A and 1 mol of B that react produce 2 moles of C. With this reasoning, if x mol of A and x mol of B react, then $2x$ moles of C are produced. Since A and B are reactants, their concentrations should be depleted as the reaction proceeds. This "withdrawal" can be entered in the checkbook in the Δ row as $-x$. Similarly, the "deposits" of $+2x$ to the C column can be entered, as shown below:

Species	A	B	2 C
initial	0.500 M	0.500 M	0 M
Δ	$-x$	$-x$	$+2x$
equilibrium			

The last step is to add the activity in each account. This "bottom line" is the equilibrium concentration as a function of initial concentrations and x. By adding the initial and Δ entries together for each product and reactant we find:

Species	A	B	2 C
initial	0.500 M	0.500 M	0 M
Δ	$-x$	$-x$	$+2x$
equilibrium	0.500 M $- x$	0.500 M $- x$	$2x$

The entries in the bottom row are the equilibrium concentrations as a function of the initial concentration and x. Since we want to know the equilibrium concentrations as a single number rather than in terms of initial concentrations and x, we must solve for the value of x. Once we know x, we can directly express [A], [B], and [C].

The solution lies in the use of the equilibrium expression. By substituting the data into the equilibrium expression, we can write:

$$0.0100 = \frac{[2x]^2}{[0.500-x][0.500-x]}$$

Note that we have one equation and one unknown. To solve for x, we must employ our algebraic skills. For this problem, the solution lies in noting that the equation can be simplified to:

$$0.0100 = \frac{[2x]^2}{[0.500-x]^2}$$

If we now take the square root of both sides, the x^2 term can be reduced to x, thereby making the equation much easier to solve.

$$\sqrt{0.0100} = \sqrt{\frac{[2x]^2}{[0.500-x]^2}}$$

which simplifies to:

$$0.100 = \frac{[2x]}{[0.500-x]}$$

By cross multiplication and rearrangement, the following equations show the solution for x:

$$0.100 \times [0.500-x] = [2x]$$

$$0.0500 - 0.100x = 2x$$

$$0.0500 = 2.100x$$

$$x = \frac{0.0500}{2.100}$$

$$x = 0.024$$

Having found x, the equilibrium values of A, B, and C can be calculated.

$$[A] = (0.500 - 0.024) = 0.476 \text{ mol/L remaining of A}$$

$$[B] = (0.500 - 0.024) = 0.476 \text{ mol/L remaining of B}$$

$$[C] = (2 \times 0.024) = 0.048 \text{ mol/L produced of C}$$

These values are the equilibrium concentrations of the reactants and products.

The use of the checkbook will allow you to setup and solve every equilibrium problem that is discussed in your text. Not all problems will be solved by the same exact algebraic methods shown above, but the solutions are similar. Example 15.7, Exercise 15.7(a) and Problems 15.72, 15.75, and 15.76 can be worked using this method. Additional chapter problems include 15.97-15.100. Note that Exercise 15.7(b) deals with the calculation of K_{eq} for the reaction presented in Example 15.7

Section 8

Skills for Chapter 16: Acids and Bases

Goals: Calculations of acid-base titrations and pH.
Skills: Multiplication, division, and logarithms.

8.1 pH Calculations

Your text discusses the concept of pH in Chapter 16.9. The unit pH is a most important term when dealing with acid-base chemistry because it describes to us the amount of H^+ ions that are present (and potentially reactive) in an aqueous solution.

When HCl is added to H_2O, the following reaction occurs:

$$HCl_{(aq)} + H_2O_{(l)} \leftrightarrow H_3O^+{(aq)} + Cl^-{(aq)}$$

Instead of writing $H_3O^+{(aq)}$ as one of the products of this reaction, we simply write $[H^+]$. As your text mentions, the production and concentration of $[H^+]$ is controlled by the dissociation of water. Water is capable of producing $[H^+]$ according to:

$$H_2O_{(aq)} \leftrightarrow H^+{_{(aq)}} + OH^-{(aq)} \qquad K_w = 1.0 \times 10^{-14}$$

A solution consisting only of water will have a H^+ concentration of 1×10^{-7} M. If a strong acid such as HCl is added to the water, the $[H^+]$ will increase.

Instead of always trying to remember the hydrogen ion concentration of an acid solution as a value expressed to a power, chemists have simplified the values through the concept of pH. By definition:

$$pH = -\log[H^+]$$

If you know the H⁺ equilibrium concentration, the pH may be obtained by entering the value of [H⁺] into your calculator, pressing the [*log*] key, and then the [+/-] to make the value positive. For example, if the [H⁺] is found to be 3.0×10^{-5} M, then the pH is calculated as:

$$\text{pH} = -\log [3.0 \times 10^{-5}]$$
$$\text{pH} = -(-4.52)$$
$$\text{pH} = 4.52$$

Please note that there are two types of logarithmic functions on your calculator, [*log*] and [*ln*]. The pH definition is based on *log* and not *ln*.

We can check the pH calculations with a given [H⁺] by using the *log* relationship of powers to pH.

$$[\text{H}^+] = 1.0 \times 10^{-3}$$
$$\text{pH} = -\log [1.0 \times 10^{-3}]$$
$$\text{pH} = -(-3.00)$$
$$\text{pH} = 3.00$$

By similar reasoning, if [H⁺] = 1.0×10^{-5}, then the pH = 5.00. For the first example worked in this section where [H⁺] = 3.0×10^{-5} M, we can predict that the pH should lie between 4 and 5. This second check with our "cerebral calculator" can be of aid in confirming the answer from our calculator. Similar problems are shown in Examples 16.10 and 16.11, Exercises 16.10(a) and 16.11(a,b), and Problems 16.53-16.54 of your text.

Another calculation that will be used in these studies will be the determination of the [H⁺] if we are given the pH. This calculation can be performed by using the concept of *antilogs*. From our earlier definition,

$$pH = -\log[H^+]$$

$$-pH = \log[H^+] \quad \text{(multiplying both sides by -1)}$$

$$10^{-pH} = 10^{\log[H^+]} \quad \text{(using the } \textit{antilog}\text{ relationship)}$$

$$10^{-pH} = [H^+]$$

$$[H^+] = 10^{-pH}$$

To perform these calculations, use the [y^x] or the [10^x] function on your calculator. If the pH is 5.5, then:

$$[H^+] = 10^{-5.5}$$
$$[H^+] = 3.2 \times 10^{-6} \text{ M}$$

A similar example is shown in Example 16.13 and Exercise 16.13 of your textbook. Problems 16.57 and 16.58 reinforce this concept.

Note that several problems involve the calculation of the pH if your are given the OH⁻ concentration. The relationship between H⁺ and OH⁻ is discussed in Chapter 16.8 of your textbook and reviewed in Example 16.12 and Exercise 16.12. From this relationship, we know that:

$$pK_w = [H^+] \times [OH^-]$$

and with the K_w of water at 25°C being 1×10^{-14}

$$14 = [H^+] \times [OH^-]$$

By knowing the [OH⁻], you can calculate the [H⁺] and then the pH. See Exercise 16.10(b) and Problems 16.53(b,d), 16.54(b,d), 16.55(c,d), and 16.56(c,d),

Of course, you can also use the mathematics introduced in this chapter to calculate the pOH and relate it to the pK_w and the pH. By defining the pOH as the -log[OH⁻], we can write the following expression:

$$pK_w = pH + pOH$$

$$14 = pH + pOH$$

This expression gives us the flexibility to calculate the pOH of a solution (when we are given the concentration of a base, such as NaOH), and translate it to the pH. Review Problems 16.51 and 16.52 for more work with pOH and pH.

An important consideration of acid-base chemistry involves the strength of the compound. This relative strength of an acid or base can be described through the concept of electrolytes (See Chapter 8.6 of your text). For instance, the acid HCl is classified as a strong electrolyte, and hence, a strong acid. The acid will completely dissociate in water; one mole of HCl yields one mole of H^+ and one mole of Cl^-. The presence of these ions in water causes the solution to conduct electricity. Therefore, it is called an electrolyte.

Other compounds may dissociate only to a small degree. If a mole of acetic acid is added to water, only about 5% of the acid will dissociate. This limited dissociation does not allow enough ions to be present in the solution in order for the solution conduct electricity. Hence, acetic acid is called a weak electrolyte.

When you are calculating the pH of an acid or base solution, you will need to determine whether or not the acid (or base) involved in the reaction is a strong or weak species. As mentioned in your text, the $[H^+]$ for strong acids is usually the initial concentration of the reactants (*i.e.* HCl) because of complete dissociation, whereas the $[H^+]$ for weak acids is determined from equilibrium calculations (as discussed in the next chapter of your text).

8.2 Titration

Another calculation for solutions that involves cross multiplication and division is titration. As discussed in Chapter 16.12 of your textbook, this process involves reacting a standard solution with an unknown solution for the purpose of determining the concentration of the unknown solution. In the lab, the commonly used titrations involve acid-base reactions, precipitation reactions, redox reactions, and complexation reactions.

Before any titration can be carried out, the solution stoichiometry must be known. Specifically, we must know how the titrant (standard solution) will react with the unknown. Using a simple acid-base reaction that is listed below, the mathematics for a titration can be set up.

$$HCl(aq) + NaOH(aq) \rightarrow H_2O(l) + NaCl(aq)$$

In this reaction, one mole of HCl will reacts with one mole of NaOH. This relationship allows the following equation to be written:

$$moles_{acid} = moles_{base}$$

Substituting the expression of mole = M × V, we obtain:

$$M_{acid} \times V_{acid} = M_{base} \times V_{base}$$

Consider a titration of an unknown NaOH solution with a standardized HCl solution. To 50.0 mL of the unknown contained in a flask, approximately 33.8 mL of a 0.175 M standardized HCl solution was added to the flask to reach the equivalence point. From this information, we can determine the molarity of the unknown solution.

If we rearrange the above equation by cross multiplication and division, we can solve for the molarity of the unknown base, M_{base}. Doing so yields:

$$M_{base} = \frac{M_{acid} \times V_{acid}}{V_{base}}$$

Putting in the data (with volumes in L), we find:

$$M_{base} = \frac{0.175 \text{ mol/L} \times 0.0338 \text{ L}}{0.0500 \text{ L}}$$

$$M_{base} = 0.118 \text{ M}$$

Compare this problem to Exercise 16.18 and Problems 16.73 and 16.74 of your text.

For an example where the solution stoichiometry is different, see Example 16.18 of your text. In this exercise H_2SO_4 is used. This diprotic acid reacts with 2 moles of the base KOH. The solution to a titration problem with this stoichiometry requires starting out with the equation:

$$1 \text{ mole}_{acid} = 2 \text{ mole}_{base}$$

The equation can be expanded to include the M and V of the acid and the base.

$$1 \left(M \times V\right)_{acid} = 2 \left(M \times V\right)_{base}$$

The information in the problem allows problem the M_{acid} to determined from the volume of acid titrated, the volume of base used in the titration, and the molarity of the NaOH solution. Compare this example to Problems 16.75 and 16.76 of your text.

Section 9

Skills for Chapter 18: Fundamentals of Nuclear Chemistry

Goals: To determine the half-life of a radioactive element.
Skills: Multiplication, division, roots, and scientific notation.

In this chapter you will be introduced to problems that concern the decay of radioactive elements. The primary calculation of these examples concerns the half-life of a radioactive element. This is described in detail in Chapter 18.2 of your textbook and is shown in Example 18.2 and Exercise 18.2

The mathematics of this example involve multiplying the original amount of a radionuclide by a fraction in order to find the amount of undecayed radionuclide. This fraction is defined as $1/2^n$, where n is the number of half-lives involved in the decay. It is written as:

$$\text{amount undecayed} = \text{original amount} \times \frac{1}{2^n}$$

The problem presented in Example 18.2 asks us to determine the fraction of a radioactive compound that would remain after a period of 6 half-lives. By setting n=6, the fraction is 1/36, or 0.0278. For every 1.000 g of radioactive material originally contained in the sample, only 1/36 g (that is 0.0278 g) would remain undecayed. See Problems 18.25, 18.26, 18.33, and 18.34 for more practice with radioactive decay.

A second use of this equation allows us to estimate the time required for a radioactive compound to decay to a prescribed level. For instance if the half-life of Bi-214 is 19.7 min, how long will it take for only 25% of the material to remain? If the original amount of Bi-214 was 1.00 g, we are trying to determine when only 0.25 g will remain. Using the equation above, we write:

$$0.250 \text{ g} = 1.000 \text{ g} \times \frac{1}{2^n}$$

$$\frac{0.250 \text{ g}}{1.000 \text{ g}} = \frac{1}{2^n} \qquad \text{dividing terms}$$

$$0.25 = \frac{1}{2^n} \qquad \text{simplifying}$$

$$\frac{1}{0.25} = 2^n \qquad \text{inverting}$$

$$4 = 2^n \qquad \text{simplifying}$$

$$n = 2 \text{ half - lives}$$

This time period would be 2×19.7 min, or 39.4 min for this amount of decay to occur. Only after 39.4 minutes would the amount of Bi-214 be 0.250 g or less. Compare this to Example 18.3, Exercise 18.3. and Problems 18.28-18.32 of your text.

Section 10

Self Test for Math Skills

Listed below are a series of questions designed to check your math skills. Please take no more than 30 minutes in answering them. You will find the answers in the back of this booklet.

1. How many significant figures are in the number 0.101?

2. Solve for x in the following equation with care to express the correct number of significant figures.

$$x = 12.011 + 1.00797$$

3. Solve for x in the following equation with care to express the correct number of significant figures.

$$x = \frac{1.057}{10.3}$$

4. Express the number 0.000356 in scientific notation.

5. Determine the mean, $\bar{x}$, for the following data: 25.12, 25.29, 24.95, 25.05 and 25.55.

6. Solve the equation: $x = 2.0 (\sin 35°)$.

7. Solve the equation: $0.405 = \cos(x)$.

8. Determine the value of x in the following equation.

$$(20.3) \times (0.0171) = \left(\frac{x}{60.3}\right) \times (1.35 \times 10^3)$$

9. What percentage of 15.5 is 1.25?

10. Solve the following equation for x.

$$\frac{1}{x} = \frac{4.3}{6.8}$$

11. Solve the following equation for x.

$$\frac{0.57}{0.22} = \frac{1.2}{x^2}$$

12. Calculate the value of x, where $x = (6.23 \times 1.22)^3$.

13. What is the fourth root of 17?

14. Solve the following equation for x.

$$x = 8.3 \times \left(\frac{1}{2^2} - \frac{1}{4^2}\right)$$

15. Determine the value of x in the following equation.

$$2.52 = \sqrt{\frac{1}{x}}$$

16. Calculate the value of x, where $x = \log(0.017)$.

17. Find the value of x, where $x = e^{\ln(12)}$.

18. Determine the value of x, where $1.53 = \log(x)$.

19. Solve for x in the following equation.

$$x = \frac{-1}{0.030} \times \ln\left(\frac{0.20}{0.50}\right)$$

20. Solve for x in the following equation.

$$x = \left(1.00 \times 10^3\right) e^{\left(\frac{-0.0220}{1.33}\right)}$$

Section 11

Chemistry and Writing

11.1 The Scientific Notebook

The scientific notebook is the scientist's own record of experiments performed and phenomena observed. Beginning with the first student laboratory report, there are special requirements for recording experimental results. The requirements may seem rigid at first, but they are very understandable in the light of the purposes of the notebook.

For the professional scientist, the claim to original work is found in the scientific notebook. Millions of dollars in patent rights may depend on the existence of a properly dated and authenticated scientific notebook. Many of the rules that are followed in recording data follow from this important function of the notebook. Nothing is ever erased; and incorrect reading is crossed out and the correct one written beside it. Work is recorded in a bound notebook with pages that cannot be removed or added. Every entry is dated, signed, and countersigned by the scientist in charge of the laboratory. All these rules are designed to produce a record that will constitute proof not only of what experiments were performed, but of the exact date. This is important, because if two scientists make the same discovery, the first one to do so will gain all the legal rights and most of the credit for the work. Obviously, it is more important to have a complete and original record than a perfectly neat one. A few crossed-out readings are not uncommon, and a few blots from spilled chemicals are not unheard of either. These are preferable in the laboratory notebook to a perfect page that has been copied over at a later date and no longer constitutes an authentic original record. **Under no circumstances is data to be recorded on loose paper rather than directly into the notebook!**

Another important function of the notebook is to record the procedure and observations so clearly and completely that the experiment can easily be repeated at a later date. Experiments that cannot be repeated by the same researcher or by other laboratories are soon discredited. For the student in the laboratory, complete notes are important as well. If something goes wrong, it should be possible to find the error in procedure from the experiment notes. At times, the numbers in the crossed-out data entries tell an interesting story. Occasionally an interesting and unexpected phenomenon will be observed that merits further study. Always, a complete, clear record of what has happened in the laboratory is essential.

In order to be a complete record, each experiment entered into the notebook should include certain features. The scientist's **name** and the **date** should always be entered. The **title of the experiment** being performed is an important element that is often neglected. "Chemistry Lab" is an inadequate substitute for the experiment title, which is usually readily available. Often it is useful to begin by writing the **objective**, or the purpose, of the experiment. Stating the objective clearly helps both the experimenter and the reader of the notebook to understand the experiment. A complete record of experimental **procedure** is essential, either as a step-by-step description or by a complete reference to a standard experimental procedure. If a standard procedure is given, great care must be made to note any deviations from that procedure. A list of **materials and equipment** used can be a great help in organization if it is included as a part of the experimental procedure.

Though the laboratory notebook does not have to be perfectly pristine, it is certainly desirable that it should be as organized as possible. Some time and thought spent in planning before the laboratory period begins will result in a better notebook and a more successful experiment. **The date, title, experimenter's name and objective of the experiment should be entered before the experiment begins.** If the experimental procedure that has been provided does not already give **labeled data tables** for an experiment, it is worth some time and thought to set up such tables before entering the laboratory, rather than waste time during the experiment deciding how to do so. Ample space should be provided not only for the expected data, but also for corrections and notes. Unused space can be crossed out later as necessary, though extra pages are never torn out. Sometimes only the right-hand pages of the notebook are used, leaving the other pages free for later notes or calculations. Individual research laboratories or student laboratories may have standard notebooks or forms in which to write laboratory results. All of them share the basic objective of recording in a useful way the scientist's actions, observations, and thoughts while in the laboratory.

11.2 The Scientific Report

When the scientist makes a formal written report of experiments performed in the laboratory, the report follows a generally accepted outline. Introduction, results and discussion, conclusions follow in order as separate sections and are clearly labeled. Lists of references and even the title are treated in standard ways.

The **title** of a scientific paper is seldom an occasion for creativity. Titles for articles in scientific journals are carefully constructed from words that will be useful key words for information searches by computer. Titles for student laboratory reports are usually indicated in the assignment. As with the laboratory notebook, "Chemistry Laboratory" is unacceptably vague as a laboratory report title. Abbreviations as part of a title should be avoided.

The **Introduction** section should make clear to the reader the purpose and the background of the experiment. The objective of the work that is being discussed should always be clearly stated. It may be appropriate to discuss the basis of the experimental methods that were used as well as the scientific theory on which the work is based. Usually a well-written introduction makes use of written resources in the form of scientific books and papers, which must be listed in the references cited and footnoted with the appropriate reference.

The **Experimental Procedure** section explains in detail exactly how the experiment was conducted. It should be possible to reproduce the experiment using the information found in this section. If standard procedures are used and not explained in detail, a reference should be given. A list of materials and equipment is often a useful component in this section. It includes all chemicals used, including the concentrations of solutions, and all special equipment.

The **Results and Discussion** section includes the data that were obtained in the experiment together with an explanation of the data. Often it is useful to organize the results of the experiment in tables, and sometimes graphs are required as well. All tables and figures should be titled and numbered. All columns in tables and both axes of a graph should be carefully labeled, not omitting units. If calculations have been performed, the equations used should be clearly indicated and enough information about the calculations should be included so that they can be clearly followed. The precision and accuracy of the results should be calculated by standard statistical methods if appropriate to the experiment.

The **Conclusions** section contains the thoughts of the experimenter about the significance of the work performed. Each part of the experiment should be discussed. Numerical results should be evaluated, and the meaning of any statistical calculations explained. The success of the experiment should be evaluated by referring to the objective of the experiment as presented in the introduction. Was the experiment successful? Were the objectives met? What is the overall significance of the experiment?

The **Literature Cited** section lists all the references used in preparing the report. This section is most formalized of all in its format. Each scientific journal has a slightly different style which contributors must follow to the letter. Student reports may also be required to follow a certain form. The best way to write this section is with the help of an example. Often college courses use scientific journals as models. The *Journal of Chemical Education, Analytical Chemistry* and the *Journal of the American Chemical Society* are examples of chemical journals which have been used in this way; the *Journal of Organic Chemistry* is often used in organic chemistry courses. When giving references it is important to notice carefully all words that are set in italics or boldface in the example references. Typesetters use different fonts for italics and boldface that are difficult to reproduce when typing or handwriting, though many word processing programs are able to reproduce them. Words that are set in **italics can be indicated by an underline. Boldface can be represented by a wavy underline.** Typically, a reference to a book appears as follows:

REID, R. C.; SHERWOOD, T. K.; PRAUSNITZ, J. M. *Properties of Gases and Liquids*; McGraw Hill: New York, 1977.

A reference to a scientific journal follows this general form:

LEE, L. G.; WHITESIDES, G. M. *J. Am. Chem. Soc.* **1985**, *107*, 6999.

11.3 Technical Writing

Scientific writing is not limited to scientific journal articles. Scientists on every level are more likely to achieve success if they are able to describe their work and explain its significance to others. Technical writing can vary from a brief explanation of how to use a piece of equipment to a lengthy report on the activities in the laboratory. Technical writers produce articles written for the layman explaining technical subjects in understandable terms. Effective technical writing is a job skill that is very much in demand. College-level assignments that involve report writing on technical subjects require the same considerations as professional writing.

First, consider the audience. Will a skilled professional or a layman read the material? If it cannot be assumed that the reader is familiar with the basic principles of the field being discussed, then the writing must include some basic background information, with special attention given to explaining technical vocabulary that may not be understood by the reader.

Most writing projects begin with a visit to the library to find appropriate source materials. Again, the level of the project will determine how the literature search is conducted. The original research reports contained in scientific journals can be found through indexes such as those provided in *Chemical Abstracts*; using the abstract indexes is a skill that must be developed through practice. Chemistry students usually are given a special course in the chemical literature that includes training in the use of *Chemical Abstracts*. Many science reports, however, require only limited use of original research papers. Science encyclopedias and dictionaries, along with periodicals written for the layman, can provide the background information for a science report and may indicate as well the authors and topics that might be explored in a more detailed technical search. Science and technology encyclopedias useful as sources of background information include:

>*Harper Encyclopedia of Science*
>*McGraw-Hill Encyclopedia of Science and Technology*
>*Van Nostrand's Scientific Encyclopedia*

More specialized encyclopedias include:

>*Encyclopedia of Chemical Technology*
>*Encyclopedia of Physics*
>*McGraw-Hill Encyclopedia of Energy*

Dictionaries can be useful in defining technical terms and concepts. Those that are useful in chemistry-related topics include:

>*Chamber's Dictionary of Science and Technology (McGraw-Hill)*
>*Chemist's Dictionary (Van Nostrand)*
>*Hanckh's Chemical Dictionary (McGraw-Hill)*
>*McGraw-Hill Dictionary of Scientific and Technical Terms.*

Facts and data can be found through the many scientific handbooks. Some of the handbooks used in researching chemistry papers include:

>*CRC Handbook of Biochemistry*
>*CRC Handbook of Chemistry and Physics*
>*CRC Handbook of Environmental Control*
>*Merck Index*

Review articles in periodicals like *Scientific American* give useful information on a variety of scientific topics. They can be conveniently found through the *General Science Index*, which provides a comprehensive subject index to English language periodical literature in the sciences. On-line computer search services are increasingly used to locate periodical references. A major resource of the library not to be neglected is the expertise of a good science librarian.

Technical writing depends no less than any other form of writing on the basic language skills of the writer. Incorrect spelling and grammar can mar the effect of the most interesting and original narrative. A good guide to English usage belongs next to a dictionary on the writer's desk. Good writing style is developed through practice in writing and rewriting. A clear, direct style contains no unnecessary words. Consider the following example:

> At this point in the experiment the mixture was heated up through the use of a hotplate.

A much-improved version is:

> The mixture was heated with a hotplate.

Some science publications prefer that use of the first person ("I heated the mixture") be avoided. Use of the passive voice "the mixture was heated" is then indicated. In other uses the more direct form of the active voice may be preferred, as in "We decided to heat the mixture" rather than "It was decided that the mixture should be heated." When writing instructions the imperative is often a good choice: "Heat the mixture on a hot plate" is more direct than "The mixture should be heated on a hot plate."

There are many references available to you to help you develop the valuable skill of communicating information. General references include:

> W. STRUNK, Jr.; E. B. WHITE. *The Elements of Style*, MacMillan: New York, 1979.

> MARGARET SHERTZER. *The Elements of Grammar*, MacMillan: New York, 1986.

References pertaining to technical information are:

> B. EDWARD CAIN. *The Basics of Technical Communicating*; American Chemical Society: Washington, DC, 1988.
>
> ANNE EISENBERG. *Writing Well for the Technical Professions*; Harper and Row: New York, 1989.

11.4 A Notebook Example

The following three pages illustrate the proper manner in which a laboratory notebook should be kept. These pages are reproductions of a student's analytical chemistry laboratory notebook. The style of this notebook conforms to the guidelines presented in Section 11.1 of this booklet.

Before entering the laboratory the student had written the introduction and experimental section. A section of data/results was begun as the student collected information. (Note that the student recorded the weights of several tablets as well as the time required to complete the coulometric titration of the ascorbic acid tablets.) This entry is reproduced on page 50.

Next, the student was required to perform several calculations with these data. They were begun in the laboratory and are found on next page (p. 51). Note that this page is the reverse page of the data entry. It would be the left-hand page of the notebook. Normally, the right hand side of the notebook is used for data entry and analysis, while the left-hand side is used for "scratch" calculations. This method eliminates the need for "loose" paper for initial calculations and other scribbles.

The last page (p. 52) contains the conclusions. Note that the student corrected the calculated values and his grammar by drawing a line through the unwanted entry. Erasure or correction fluid was not used.

This simple system of using the different sides of the notebook for formal entries and simple calculations can greatly enhance your ability to organize your data and to present your findings.

Coulometric Titration of Ascorbic Acid with Iodine   3/23/00

Introduction: The purpose of this experiment is to determine the weight percent of ascorbic acid in a vitamin tablet.

Experimental: Add 0.1M KI solution and a spatula of soluble starch. Cautiously add 1 sec increments to the current until a pale blue color appears. This will be the matching color. Crush a vitamin tablet and weigh 20 to 30 mg to the nearest 0.1 mg; transfer to the cell. Turn on current. Repeat the procedure with 2 more samples from the tablet. Continue to analyze other tablets.

Data/Results: multiplier 0.05 = 48.25 mA
250 mg tablets / 361 mg tablet #1, 364.8 mg tablet #2

sample	weight	time
1	0.0232 g tablet	389.2 sec
2	0.0225 g	368.4 sec
3	0.0243 g	————
4	0.0249 g	385.0
5	0.0240 g	359.1
6	0.0234 g	353.0
7	0.0240 g	366.3

Actual Weight	Time	Ascorbic Acid		Weight %
0.0232 g	389.2 sec	0.0171 g	Tablet 1	73.7
0.0225 g	368.4 sec	0.0162 g		72.0
0.0249 g	385.0 sec	0.0170 g		68.3
0.0240 g	389.1 sec	0.0158 g	Tablet 2	65.8
0.0234 g	353.0 sec	0.0155 g		66.2
0.0240 g	366.3 sec	0.0161 g		67.1

Tablet #1 contained on average 263.0 ± 4.3 mg ascorbic acid.
Tablet #2 contained on average 244., ± 4.2 mg ascorbic acid.

263.0 ± 4.3
244.1 ± 4.2

$$95\% \; CL = \bar{X} \pm \frac{ts}{\sqrt{N}}$$

$$= 263.0 \pm \frac{12.7(4.3)}{\sqrt{2}} \qquad = 244.1 \pm \frac{3.18(4.2)}{\sqrt{4}}$$

$$= 263 \pm 38 \qquad\qquad = 244.1 \pm 9_{\;6.7}$$

$\bar{X} - \bar{\mu} = 263 - 250$ $\qquad \frac{ts}{\sqrt{N}} = \frac{(12.7)(4.3)}{\sqrt{2}}$
$\qquad = +13$ $\qquad\qquad\qquad\qquad = \pm 38$

$\bar{X} - \bar{\mu} = 244 - 250$ $\qquad \frac{ts}{\sqrt{N}} = 6.7$
$\qquad = 6$

Conclusion: The value given on the box for each ascorbic acid tablet was 250mg. We The obtained results for two tablets were 263.6 ± 4.3 mg and 244.1 ± 4.2 mg. The results are close to the reported value. Some error can be expected in this experiment because the calculations depend on an observed endpoint, and although it is tried, it is sometimes hard to pick the same endpoint each time based on color.

The third run on the first tablet was discarded because the electrode was not in the solution and therefore no reaction was taking place.

The 95% CI produces results of 263 ± 38 for the first tablet, which rounds to 263 ± 38. This is mainly high because it is based on two readings. The second tablet gave a result of 244.1 ± 6.7 mg. This is based on four readings.

A t-test on the first tablet gives $+38 < +13$, which says that our result is good and would, on average, be correct 95 of 100 times. Our second tablet gives $-6.7 < 6$, which suggests that our second value can also be accepted.

The overall reliability of Coulometric titrations is good as long as you pick an endpoint and can reproduce it.

Section 12: Careers in Chemistry

Chemists often find themselves in careers that seem unrelated to traditional chemistry, yet draw heavily on chemical knowledge and skills. Some chemists go into management; management positions in science-related industries increased by 30% between the years 1994 and 2000. Technical sales, patent law, and forensic science are only a few of the careers open to holders of chemistry degrees. Laboratory workers are needed in materials science, polymers, and biotechnology; all these fields depend on the molecular science of chemistry. The traditional skills of chemical analysis and synthesis are in demand for such areas as environmental testing and pharmaceutical development. A wide variety of rewarding career options is available for those with chemical training. The overall unemployment rate for chemists is very low, averaging about 1%.

12.1 Materials Science

One of the major predicted growth areas in the economy is in the area of materials science. The new superconducting materials that promise major breakthrough applications in fields as diverse as communications and transportation are the products of materials science. The development of new graphite materials is resulting in new types of tennis rackets, aircraft, and auto bodies. New ceramic materials are being researched for a variety of uses, including automotive pollution control. Fiber optic materials are revolutionizing communications.

Like many of the fields that offer exciting new job opportunities, materials science is interdisciplinary, requiring knowledge of chemistry and physics. The field employs both chemists and chemical engineers. Courses in polymer science, metallurgy, and computer science are also helpful. Industry employs the largest number of materials scientists: government and university positions are possible in this field as well.

12.2 Polymer Science

The long-chain giant molecules called polymers form a special class of materials that find a wide variety of uses. Synthetic fibers like nylons and polyesters are the basis of a major industry. Packaging materials as diverse as Saran Wrap and Styrofoam are made of polymers. Televisions, computers, toys, tapes, and CDs all make use of polymeric materials. Increasing concern about solid waste disposal is prompting research on plastic recycling and degradability. A promising research area involves modification of polymer properties to make materials that will be compatible with human tissues for medical transplants. Preparation for an industrial career in polymer science may involve a degree in chemistry, chemical engineering, or polymer science and engineering.

12.3 Environmental Science and Technology

Increasing concern about the environment has created numerous job openings for scientists who want to find solutions to the problems caused by pollution. Understanding the chemical and biochemical reactions that produce and consume chemicals like carbon dioxide and methane in the atmosphere is critical to understanding the possible long-range warming process known as the greenhouse effect. Chemists discovered that the chemical reactions of CFC's, or compounds containing carbon, fluorine, and chlorine, with ozone in the upper atmosphere, were causing ozone depletion and a resultant increase in the amount of harmful ultraviolet radiation that is reaching the Earth's surface. Now chemists must discover substitutes for these compounds that will replace them in such uses as refrigerants for refrigerators and air conditioners. Automobiles and factories must be designed so that their combustion processes do not foul the air. Factory effluent must be treated or recycled so that streams and groundwater are not polluted.

State and federal agencies such as the Environmental Protection Agency employ chemists to monitor pollution and help to find ways to decrease its sources. Waste management companies recycle materials or dispose of them responsibly, an increasingly technical task. Chemists, biochemists, and chemical engineers all find employment in areas of environmental technology.

12.4 Biochemistry and Biotechnology

Biochemists study the chemistry of living systems. Understanding the reactions which occur in the human body, in animals, plants, insects, viruses, and microorganisms makes new approaches to curing disease and improving food technology possible. New information from the human genome project will have profound implications for our society. Colleges and universities employ almost half the biochemistry workforce; the rest are employed by government agencies or private companies.

Biotechnology is a burgeoning new interdisciplinary field employing biochemists, chemists, and biologists. Most of these are employed in industry, either by established pharmaceutical and agricultural companies or by new biotechnology venture firms in this rapidly growing area. Through biotechnology, scientific breakthroughs in molecular biology are used to develop new products for the commercial marketplace. Microorganisms are being produced on a large scale which are able to produce insulin or the human growth hormone. Specific microorganisms are being produced with the goal of consuming oil spills or hazardous chemical waste. Genetic variation of plants may produce varieties, which are richer in nutrients or resistant to insects. These are but the beginnings of the commercial applications, which may be expected in biotechnology.

12.5 Medicinal Chemistry and Clinical Chemistry

Biochemistry and biotechnology are improving both our understanding of diseases and the methods of treatment, but they are not the only fields of chemistry related to medicine. Medicinal chemistry and clinical chemistry are specialties in which chemists have a direct impact on health care.

The medicinal chemist develops new therapeutic agents. Chemical compounds are designed and synthesized with the aim of producing molecules, which will act on one area of the body, such as a certain part of the brain. Once the compound has been synthesized, it is ready for the long series of tests for both positive and harmful effects, which all new drugs must undergo. Tests may indicate that further modification of the molecule's structure is required. To prepare for this work, which links chemistry, biology, and medicine, requires training in chemistry, pharmacology, or medicinal chemistry.

The clinical chemist applies the techniques of analytical chemistry to body fluids. A variety of chemical instrumentation is used, and the modern clinical laboratory makes extensive use of computer automated testing. Administrative duties may form an important part of the clinical chemist's work. Clinical tests are performed both to monitor normal body functions and to test for therapeutic and toxic drug levels in the body. Often critical decisions about health care are based on laboratory results. Many clinical laboratories are involved in research-related testing, to find the results of new procedures, drugs, and equipment or to support basic research. Clinical chemists are employed by hospitals, universities, government and industry.

12.6 Forensic Science

Forensic chemists apply the skills of analytical chemistry in the crime laboratory. Suspected samples of drugs must by analyzed for authenticity by chemical instrumentation if they are used as legal evidence. Body fluids and samples of body tissue are provided by the medical examiner for analysis in the case of a homicide, and cases of sexual assault involve laboratory testing as well. Traces of blood or of gunshot residue can be revealed by chemical tests. Other skills are required of the general forensic scientist, such as investigation of the crime scene. Though these are not directly related to chemistry, some of the most eminent forensic scientists were trained as chemists. Most forensic scientists are employed by government laboratories.

12.7 Radiochemistry

An exciting area of chemistry which is experiencing severe shortages of trained workers is radiochemistry. This specialty is needed in a multitude of application areas, some of which are growing rapidly.

Nuclear medicine is seeing rapid advances in such areas as the use of monoclonal antibodies in cancer research. The radiopharmaceutical industry needs radiochemists both to manufacture radionuclides and to oversee waste disposal and protection from radiation hazard. Positron emission tomography (PET) is a diagnostic application of radiopharmacology, which requires radiochemists to staff PET centers and to develop and provide radionuclides and labeled compounds. PET is expected to expand rapidly as its use shifts from primarily research use to more routine medical applications.

Nuclear power production requires radiochemists both in power plants and in related industries to oversee the running of plants and to supervise waste treatment.

Environmental chemistry makes use of radiochemistry in several ways. Accidental emissions from power plants such as the Chernobyl accident require skilled monitoring to assess environmental and health effects. Radionuclides can be used as tags to monitor complex systems such as the buildup of greenhouse gases or the sources of atmospheric CFC's,, which destroy the ozone layer. Neutron activation analysis can analyze for trace elements in rocks and help explain Earth's history.

12.8 Patent Attorneys and Patent Agents

A background in chemistry can serve as preparation for the career of patent attorney or patent agent. A patent attorney must have a law degree as well as a science degree. A patent agent may have a degree in chemistry, physics, engineering, or related technological fields. For certification, the patent agent must pass an examination on patent procedures and rules. The patent agent examines the patent literature to determine whether a client's invention is patentable and writes patent applications. Patent lawyers may in addition represent a client in litigation on licensing, trademark or copyright issues, trade secrets cases, and antitrust cases. Patent agents and attorneys are always well paid and in short supply. Their work involves communication skills as well as scientific training. They are in demand both by government offices and private law firms.

12.9 Chemical Education

Chemistry is taught at both the high school and the college level. College chemistry is taught at two-year and four-year colleges and at universities. Teaching positions at larger universities are likely to involve greater emphasis on graduate education and research than on undergraduate instruction. Demand for high-school teachers is increasing in most areas of the country.

Teachers at all levels speak of the satisfactions involved in interacting with young minds and influencing future careers. Individual freedom is often greater than in many other types of employment, and scheduling is more flexible. College professors have many obligations in addition to teaching, however. Committee responsibilities are required as a part of participation in college governance, and administrative duties may be part of some professors' workloads. Research involves not only study and planning, but also the training of graduate students and postdoctoral fellows and the writing of proposals for research funding. Financial rewards may not be as great as those of the industrial chemist, especially at the high school level and at smaller higher education institutions. For those who teach, the quality of life in the work experience is often a major part of the reward.

12.10 Chemical Information Careers

Chemical information is produced at an increasingly rapid rate as new discoveries are made and the results published in scientific journals. Several career possibilities are available to the chemist who can help others to locate and understand the relevant chemical information for their needs.

The chemical information specialist is familiar with both the chemical literature and the on-line computer services through which chemical information is increasingly accessed. Libraries use chemical information specialists, but chemical companies need them also to help their researchers. Often toxicological and environmental information is needed as well as experimental data retrieval. This special kind of problem-solving expertise should remain in demand as information retrieval becomes an increasingly complex field.

Abstracting services summarize and index chemical discoveries so they can be found by their researchers. They employ chemists to prepare information both for written abstract material and for on-line search services.

Science writers are needed on all levels, both to edit and write scientific publications for the scientists and to interpret scientific information in terms the general public can understand. Research organizations, medical centers, technical companies, and government agencies all need writers who can communicate clearly with a variety of audiences. In addition to print media, radio and television employ experts in science communication.

12.11 Other Chemistry Careers

This list of some current options in chemistry careers is by no means complete. Chemists work in museums and metallurgy plants, in food science laboratories and consumer testing services, in adhesives research and photography laboratories. The materials of our modern world are developed by chemists, and chemists monitor the safety of these products and of our environment. The materials of our future and discoveries, which will prolong our lives and improve the quality of life will be the products of chemical research.

Section 13

Study Aids for Chemistry

Chemistry is a field of science whose basic foundations and concepts goes back for many centuries, if not a millennia. All of this activity by scientists has produced a great body of knowledge, which is collectively called "chemistry". The vastness of this body of knowledge is more than one individual could ever know. Even with a lifetime of study, no one person could ever be well learned in all the areas of chemistry.

However, the study that you are to embark on this year with the text **Fundamentals of Chemistry**, is a well organized guide into the basic concepts of chemistry and the connectivity that chemistry has will every facet of modern life. Over the coming year, you will be exposed to discoveries and wonders in chemistry that now define our modern life and offer ways to improve the quality of life for all people.

Having said this, let us emphasis that the material presented in this course will not require all of your waking hours to successfully comprehend and master. You have probably heard "horror stories" from older students concerning chemistry. As chemistry teacher (and one who has taught thousands of students), let me say that their stories are quite overrated. It has been my perception that the students who do not do well in chemistry are often the ones that have not learned how to study effectively.

The habit of good studying is not one that comes naturally. It is a learned process. It is not a hard process; in fact most of the basis of these habits are founded in good "common-sense".

On the next few pages, you will find suggestions on how to develop good study habits. Also listed are suggested ways to get the most out of your textbook, ways to relax while studying chemistry, and tips on taking exams. You may already be practicing many of these techniques. If you see new ones, try and incorporate them into your study routine.

Life is a learning process. While the focus of our learning will change as we go from school to a job, the habit of good studying really does not change. Good study habits can make you a "quick-learner".

Most of all, good study habits can help you enjoy chemistry. That's right, enjoy chemistry! It is truly filled with wonder and amazement. It is rich in history and is relevant to your everyday life. The understanding of the concepts presented in your text and reinforced in this booklet will most certainly allow you to see the connectivity of chemistry to all that is around you. While good study habits will allow you to go beyond the memorization of the material and into critical thinking, it is the skill of critical thinking that sets the stage for your enjoyment of chemistry. This enjoyment from your understanding of chemistry can be truly wondrous.

13.1 Good Study Habits

Good study habits are not among our "basic instincts." Most of us have to work long and hard to develop good study habits. Here are some concrete things you can do to promote your success as a student.

- Set a long-term goal, that is, a career goal. Goals keep us going when the going gets rough.
- Set short-term goals, that is, daily or weekly goals. Short-term goals motivate us and keep us on target.
- Make a realistic daily, weekly, and term-long schedule. Post copies of your daily schedule where others can see it. That will force you to stick to it. Also schedule relaxation time for yourself! We all need that!
- Observe "good students" and make yourself look and act like a good student. Soon you will start thinking of yourself as one. We often have to "act" ourselves into a different way of thinking.
- Create a study area that meets your needs. The study area should have few distractions and be quiet. Try to keep it uncluttered. Having "a place" to study helps you get in the mood to study.

- Select the best time for you to study. We all have different "internal clocks." Study when you are most alert.
- Reflect and review daily. Reading a lesson and doing problems is not studying. You read the lesson and do the problems to gain the knowledge you need to study the material. Studying involves moving material from memory to understanding.
- If you can't explain a concept in your own words and relate it to something you already know, you probably don't understand it and won't remember it very long.
- Get to know others in your class and form a serious study group. Then study together at least once a week. It will keep you "honest." It is easy to slip into sloppy study habits and fool yourself into thinking you know more than you do. Having to explain ideas in a study group will bring you back to reality very quickly. We learn and learn how to learn from working with others.
- Work to stay relaxed while you are studying.

13.2 Reading Your Textbook

If your assignment is to read pages 88-108, the worst thing you can possibly do, aside from not doing the reading at all, is to open the text to page 88 and begin reading. Reading an assignment is a three-phase process. It has a *before*, *during* and *after* phase and each is equally important!

The *before* phase:

- Skim the entire assignment first so you have a general idea of the content and try to relate the content to concepts you already know.
- Look at your syllabus and see how many lectures are devoted to the assignment. See what other material is included with this assignment on the next test.
- Judging from the length and the difficulty of the reading assignment, try to determine how long the task will take and plan ac-

cordingly. (Remember that reading technical information often takes more time than reading non-technical works.)

The *during* phase:

- Mark up your text. Write notes in the margin. Highlight key concepts. (Try not to buy a book that has been marked up too much. You will be distracted by someone else's markings.)
- If examples do not show all the steps, write in the missing steps so you will remember how to do the problem when it is time to review for a quiz or test.
- Look at the pictures, figures and graphs in the text. Remember, Alice in Wonderland asked "What good is a book without pictures?" Fortunately most modern chemistry books are filled with pictures. Take advantage of them. Remember that chemistry is visual. When the text refers to a drawing, figure, diagram, chart, photograph, etc., look at it right away. Study it. Sometimes one picture can be worth many words!
- Spend time reflecting on what you read. Can you put the concepts into your own words?
- Make a short outline of the material you have just read. Placing new concepts into your own words helps you to retain and master the new material.
- Do you need to hear yourself reading the material? Reading difficult material aloud slows you down and lets you hear as well as see. Two senses are better than one! Remember that you are reading for understanding.

The *after* phase:

- Go back to any sections you did not understand. If after going over them again, you still have concerns, put a question mark beside them and ask about them during the next class or help session or meeting with your study group.
- Go back over the entire reading assignment looking at main headings and the notes you made in the margin. Does this information

remind you of the main ideas and content of each section? If not, add to it.
- Write a brief summary of what you read. Include any type of diagram that will help you remember the material. This summary, what you have underlined in the text, what you have written in the margin in the text, and the notes you have taken in class are the materials you need to use when you review daily and study for quizzes and tests.

13.3 Learning to Relax

It's hard to study effectively and you certainly won't perform your best on an exam if your body is overly tense. Also when you are tense, things always look worse than they are. However, some stress is normal when you enter into the unknown. And a little tension keeps you from getting too complacent and making careless errors. It is the excess tension that you need to be able to control.

There are three factors you can consciously control to reduce your stress and increase the effectiveness of your studying: posture, muscle tension and breathing. We will discuss just a few techniques you can try with each factor.

Posture

When we are concentrating hard, especially when we are tired, we often tend to pull our neck down into our shoulder blades, collapse our chest and let the lower spine bow out. Sitting with the body in this position tends to make us tired and tense. Our muscles work overtime to hold the body up and the lungs cannot breathe efficiently.

It is important to have a good posture while sitting and studying. If you cannot hold a good posture, you may need to take a break and relieve muscle tension.

Muscle Tension

A tense muscle uses more energy and requires more oxygen and nutrients than a muscle at rest. But a tense muscle constricts the capillaries that deliver oxygen and nutrients to it, which leads to fatigue of the muscle. The exhausted muscle becomes even more tense starting a downward cycle. There are three things you can do to relieve muscle tension: stretching, massage and movement.

- While you are studying, remember to stretch. Stand up periodically and stretch your body.
- Take a walk. Walking is one of the best ways to relax. Even a short walk around the block will help you relax and feel more alert.
- Periodically look out the window or across the room to relax your eyes.

Breathing

Your breathing is tied to your emotions. Have you noticed after you have been scared, that your breathing becomes rapid and shallow? This is one symptom of what is known as "fight or flight," a remnant of the days when we had to either confront a danger or run. In the modern world this reaction does little to help us.

The same "fight or flight" reaction can happen on a smaller scale when you are studying unfamiliar material or when you are taking an exam. Here are some breathing techniques you can try anytime you feel tense.

- Breathe in slowly through your nose as you count to five.
- Then breathe out slowly.
- Pause for a second when you have completely exhaled, then repeat for a few more breaths.

When you inhale, expand your abdomen first to fill the lower lungs. Then let your chest rise to fill the upper lungs. Don't force the breath; let it come easy and natural. You can do this exercise a few minutes before you take an exam.

13.4 Conquering Tests

Preparing for the Test

- Test preparation begins on the first day of the course. Keeping current with all daily assignments is the most effective method of preparing for tests.
- Form a study group. You never really learn anything until you have to explain it to someone else. A study group will force you to do this. Working with others will also help you learn more study strategies. But study on your own both before and after the study group.
- Tests are more difficult than individual assignments because they contain a greater variety of material. To avoid getting confused, you must "overlearn" the material. To "overlearn," study actively until you really feel that you understand the material or can work the problems with no difficulty. Check to see how long you had to study or how many problems you had to work to reach this point. To "overlearn," you must spend at least half as much more time or work half as many more problems. Then you must also spend a small amount of time each day, maybe 10-15 minutes, reviewing past assignments after you have completed the new assignment.
- If there is a concept you do not understand, get help immediately. In chemistry learning is often sequential, *i.e.*, new lessons build on past lessons. Therefore falling behind can be fatal!
- Try to predict test questions as you are studying.
- If practice tests are available, get them as soon as possible and do them at least several days before the test. Then you will still have

time to get help if you find there are questions that are giving you difficulty.
- If you do fall behind in your studies, never try to learn new material the day or the night before a test. You will not have time to "overlearn" it. Therefore what will normally happen is you will confuse new material with what you already know. This confusion can cause you to miss even more test questions than you would have missed had you not tried to "cram in" the new material. Concentrate on what you know on the night before the test. Then, after the test is over, catch up immediately.
- The night before the test be sure to get enough rest. Also check your calculator and put it and any other materials you will need in a place where you will be sure to pick them up before leaving for class.

Taking the Test

- Wear a watch! Bring your calculator and make sure it is functioning.
- Get to class a few minutes early so you have time to get comfortable and get organized before the test is handed out.
- Put your name on the test!
- As soon as you get the test, write down on the test paper any information you are worried about forgetting. Then you won't have to worry anymore, and your mind will be free to concentrate completely on each test question.
- Preview the entire test. It only takes a minute or two but can tell you much. Previewing tells you: the number and types of questions the test contains (multiple choice, true-false, matching, essay, problems, etc.), the number of points each question is worth, any additional information you need to write down to be sure you remember it, and how to schedule your time.
- Previewing should also build your confidence because, if you are prepared, you will know immediately that you can answer most of the questions.

- Read any directions carefully and completely. If there is something you do not understand, raise your hand and ask about it. The instructor will probably be able to clarify it for you. In any case, the worst that can happen is the instructor will say he or she cannot answer your question.
- Begin with the questions that are easiest for you. This strategy will build your confidence and lessen your anxiety. It may also remind you of things you need to know to answer the more difficult items. In addition, it will prevent you from losing points by spending too much time on things you don't know and running out of time before completing items you do know.
- Return to questions you skipped. If they are objective questions, i.e., true-false, multiple choice, matching, etc., make an "educated guess." (See *Mastering the Multiple Choice Test*). If they are essays or problems to be solved, do as much as you can. If the problem has several dependent parts and you do not know how to do the first part but do know how to do the next part, write a note saying, "I'm not sure how to do part A, so I am going to assume the answer is (make up a reasonable value) and use this value to complete the problem." Remember that part credit is better than no credit! Answer every question as completely as possible.
- Be neat and show your work, it will be easier to check your answers.
- Check your work! If you only have a few minutes left, check the first few questions you answered. You were most nervous then, and therefore you might have made careless errors. Then check the problems you thought were the easiest. You also make careless errors when you are too relaxed. If you eliminate careless errors, you can usually increase your grade by at least 10%!
- Do not get nervous because others are leaving before you. Those who finish early often are the weakest students. Use all the time allowed. There is no prize for finishing early!

Mastering the Multiple Choice Test

In addition to the general test-taking strategies presented in this booklet, there are some specific things you can do to increase your score on multiple-choice tests.

- As soon as you get the test paper, jot down any information you know you will need and are afraid you will forget.
- Scan the entire test. If any of the questions remind you of any other facts you are afraid of forgetting, write them down.
- Read the directions. Most multiple choice tests have only one correct choice, but don't assume anything.
- Answer the easiest, least time-consuming questions first. This strategy will assure that you have time to answer all the questions you know. It will also build your confidence and remind you of things that might help you answer the more difficult items.
- Before you look at the answer choices, work the problem and ask yourself "Does this answer make sense?" Remember that many of the incorrect answer choices will be derived from mistakes that students commonly make. If you see your answer as one of the choices, don't immediately mark it.
- If you don't find your answer among the choices, look again. Perhaps your answer is there but in a different form. For example 1/2 could be written as 0.5, 0.50 or 50%, etc. Maybe the answer is in scientific notation and you calculated your answer in decimal form.
- After you've answered all the questions you know, carefully re-read each question you skipped. You may find you now remember how to do them. Doing the questions you know warms you up and sometimes answering one question will spark your memory on how to answer another.
- If you are still unable to answer some of the questions, make an "educated" guess. Can you eliminate one or more of the choices? For example, if the question asks for grams, eliminate all the choices that don't have units of grams.
- Never leave a multiple-choice question blank unless the directions say there is a penalty for guessing.

- Use all the time allowed. There is no prize for being the first to finish.

Section 14

Answers to the Self Test of Math Skills

	answer	if you missed this problem, refer to:
1.	3	Section 2.1
2.	13.019	Section 2.1
3.	0.103	Section 2.1
4.	3.56×10^{-4}	Section 2.2
5.	25.19	Section 2.3
6.	1.1	Section 1.2
7.	66.1°	Section 1.2
8.	0.0155	Section 1.1 and Section 2.2
9.	8.06%	Section 2.2
10.	1.6	Section 1.1 and Section 5.1
11	0.68	Section 1.1
12	439	Section 1.2
13.	2.03	Section 1.3
14.	1.6	Section 1.3
15.	0.157	Section 1.3
16.	-1.77	Section 1.3
17.	12	Section 1.3
18.	34	Section 1.3
19.	31	Section 1.3
20.	984	Section 1.3